Geography Zone: Landforms™

PLAINS

Emma Carlson Berne

PowerKiDS
press™

New York

Published in 2008 by The Rosen Publishing Group, Inc.
29 East 21st Street, New York, NY 10010

First Edition

Editor: Joanne Randolph
Book Design: Julio Gil

Photo Credits: All images Shutterstock.com.

Library of Congress Cataloging-in-Publication Data

Berne, Emma Carlson.
  Plains / Emma Carlson Berne. — 1st ed.
      p. cm. — (Geography zone—landforms)
  Includes index.
  ISBN 978-1-4042-4204-3 (library binding)
  1. Plains—Juvenile literature. I. Title.
GB571.B47 2008
551.45'3—dc22

                                    2007031759

Manufactured in the United States of America

Contents

A plain is land that is flat. Plains do not have hills or mountains. Plains can be thousands of miles (km) long. Plains can also be only a few miles (km) long.

Plains are on every **continent** on Earth, except for Antarctica. Plains can be covered with trees or covered with grass. Plains can also be sandy, rocky deserts.

The Great Plains are huge plains in the United States, Canada, and Mexico. They are around 2,400 miles (3,860 km) long and about 1,000 miles (1,609 km) wide.

This is the Serengeti Plain, in Africa, during the wet season. Rains fall in December and from March to May, making the grass green.

Plains are part of Earth's ever-changing surface. **Glaciers**, wind, and floods form plains. Over time, new plains will form, and old plains will disappear. The Great Plains were formed around 70 million years ago. Today, wind sweeps across the plains, moving earth around and changing the shape of the land. Rivers cut through the plains, creating canyons. Rivers also move and **deposit** earth in new places.

The same kinds of things are happening in plains around the world. If we could come back in another 70 million years, who knows what we would see!

The Mississippi River makes it possible to farm these plains. Over time the river will change the way the plains look, though.

Coastal plains are on the coasts of continents, near the ocean. They are like a part of the ocean floor. The ocean floor slopes upward as it nears shore. The water gets less deep. Finally, dry land appears. This land is the coastal plain. It can reach for hundreds of miles (km) back into the continent.

Much of the United States' East Coast is a coastal plain. It is part of one of the biggest coastal plains in the world. Because it is next to the Atlantic Ocean, it is called the Atlantic Coastal Plain. Another coastal plain in the United States is the Oxnard Plain, in California.

People enjoy spending time on the beaches of coastal plains. During large storms, coastal plains can be in danger of flooding, though.

Inland plains are in the middle of continents. Europe, North America, South America, Asia, and Australia all have huge inland plains.

Glaciers made these inland plains. When glaciers pushed through the land thousands of years ago, they filled up lakes and seas with dirt and rocks. They smoothed out the hills. When the glaciers melted, the flat inland plains were left behind.

The Kalahari Desert is a giant inland plain in Africa. It is 360,000 square miles (932,396 sq km) wide. The Kalahari is covered with soft, red sand for miles and miles (km).

Here you can see the red sands of the Kalahari Desert. Around 400 known kinds of plants grow there, though most are grasses and acacia trees.

Flood plains are the flat land on either side of a river. When the river floods, the river water spills out over the banks. The water spreads over the flood plain.

Soon after, the river water sinks into the ground. It leaves behind a layer of mud and **silt**. The mud and silt make the land **fertile**. Farmers can grow crops in the rich, fertile soil on the flood plain.

The Nile River in Egypt has a fertile flood plain. For thousands of years, farmers have grown crops in the soil left behind after the Nile River floods the land.

This farmland lies on the fertile plains near the Nile River in Egypt. A man-made waterway brings water from the Nile to the farm.

Many plants and animals live on plains. On African plains, elephants and giraffes look for food and water. Lions hide in the grasses and wait for **prey**. Giant **baobab** trees spread their twisted branches over the flat, grassy land.

In North America, wolves, antelope, and deer live among the tall bluestem grasses and switchgrass of the Great Plains. Blackbirds rest on grass stems while hawks fly overhead, hunting for mice and rabbits to eat.

In the southern United States, pine trees cover sandy coastal plains. Red foxes scurry among the trees. Snakes slither along the ground, and herons fish in swamps.

Bison and elk eat grass on the Great Plains. There are about 500,000 bison on the Great Plains today.

The Atlantic Coastal Plain is a huge plain on the eastern edge of North America. This plain stretches from Cape Cod all the way to the Yucatán Peninsula, in Mexico. It is 3,200 miles (5,150 km) long and in some places, it is 300 miles (483 km) wide.

Big cities, such as Newark and Baltimore, lie on the edge of the Atlantic Coastal Plain in the United States. There are also beaches on the shore of the plain. Farmers grow crops on the Atlantic Coastal Plain, too. In Florida, they grow oranges. In South Carolina, they grow cotton.

These oranges grow well in the soil of the Atlantic Coastal Plain in Florida. The rest of America counts on Florida's oranges for juice and food.

Spread over the African countries of Tanzania and Kenya is a giant, grassy plain. It is called the Serengeti. Millions of animals live on the Serengeti. Herds of gazelles, zebras, and **wildebeests** travel the land. Cheetahs, leopards, and baboons live there, too.

The Serengeti is a special place. It is one of the only places left in the world where huge herds of plains animals **migrate** each year. In the fall, millions of zebras and wildebeests travel from the north to the south. They are looking for food and water. In the spring, they return again to their northern home.

Giraffes and zebras call the Serengeti home. Giraffes eat the leaves of acacia trees, which grow on the plain.

Many **tribes** of native people make their home on the Serengeti. Some are farmers, and some are **hunter-gatherers**. Other people raise and herd animals.

Some tribes are nomadic. These people have no fixed home. Instead, they move from place to place, following herds of animals and gathering food along the way.

The Maasai tribe lives on the Serengeti. They raise herds of cows. They eat the cows and drink their blood and milk. The Maasai sometimes do special dances in which the dancers jump high into the air over and over.

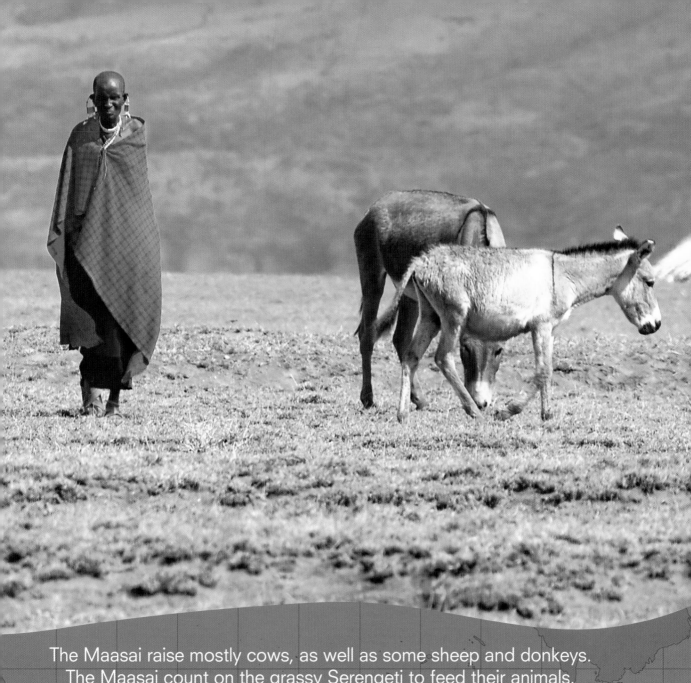

The Maasai raise mostly cows, as well as some sheep and donkeys. The Maasai count on the grassy Serengeti to feed their animals.

People all over the world like to live on plains. Because plains are flat, it is easy to build roads and towns on them. People can get from one place to another quickly on plains.

People also like to farm on plains. The land is easy to plant crops on because it is flat. Often, the soil on plains is rich and fertile.

Many factories are also on plains. It is easy to move the factory goods from one place to another across the flat plains. People use trucks and trains to carry the goods to cities.

Glossary

baobab (BOW-bab) A kind of large tree that grows in Africa and has a very thick trunk.

continent (KON-tuh-nent) One of Earth's seven large landmasses.

deposit (dih-PAH-zut) To leave behind.

fertile (FER-tul) Good for making and growing things.

glaciers (GLAY-shurz) Large masses of ice that move down a mountain or along a valley.

hunter-gatherers (hunt-er-GA-ther-erz) People who live by hunting animals and gathering wild food instead of growing crops.

migrate (MY-grayt) To move from one place to another.

prey (PRAY) An animal that is hunted by another animal for food.

silt (SILT) Fine bits of earth, smaller than sand grains, found at the bottom of lakes, rivers, and streams.

tribes (TRYBZ) Groups of people who share the same customs, language, and kin.

wildebeests (WIL-deh-beests) Large African animals with a short gray or brown coat and horns.

Index

Web Sites

Due to the changing nature of Internet links, PowerKids Press has developed an online list of Web sites related to the subject of this book. This site is updated regularly. Please use this link to access the list:
www.powerkidslinks.com/gzone/plain/